THIS BOOK BELONGS TO:
I0813546

To all the future scientists
and wildlife conservationists
out there

A TINY HABITATS BOOK

bog buddies

Amy Hevron

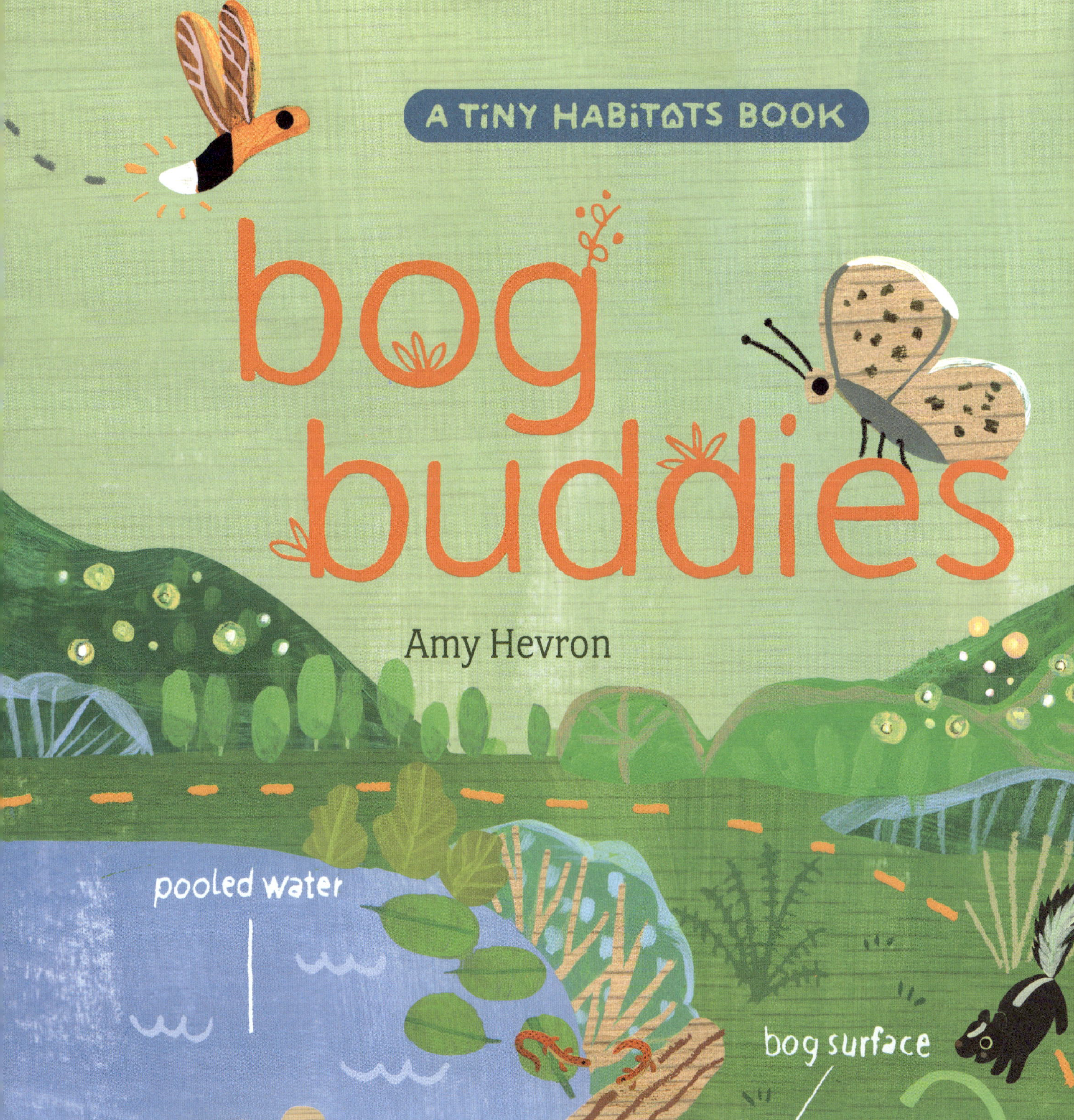

Beach Lane Books
New York Amsterdam/Antwerp London Toronto Sydney/Melbourne New Delhi

Welcome to the majestic misty mountains . . .

where fog fills alpine valleys,

rains flood roaring rivers,

streams flow down to a meadow . . .

and collect in a mossy mountain bog.

summer

month 1

Like me!

And me!

And me!

The peat mosses were the first to soak up the heavy summer rains at the bog.

They stretched out
their leafy stems and
gulped up water.
Sip!
Slurp!

Soon, a teeny-tiny bog turtle stopped by.

Wood ducks whistled.

Wood frogs frolicked.
Hi, duck!
Hi, frog!

And a water shrew met
new friends in the deep end.
Hi, breakfast!
Wait, what?!

Pitcher plants filled up on rainwater.

Beavers filled up on wet logs.

And mating fireflies filled the sky with twinkly lights.

Meanwhile, the bog soaked up summer floodwaters like a sponge, keeping the forest from flooding.

As weeks went by, the bog became a peaty playground.

autumn

month 4

Salamanders palled around in a pool.

Flying squirrels scooped up cranberry snacks.

And newly hatched firefly larvae became fast friends.

Baby bog turtles teeter-tottered among the arrowheads.

Baby snails slip-slided on lily leaves.

And beaver besties
seesawed on the
black gum trees.
Look out, little warblers!

Meanwhile, the spongy bog filtered the fall rains and filled underground springs with fresh water.

As the colder months set in, the bog got ready for a snowy sleepover.

winter
month 8

Lemmings munched on moss.
Mmm, frosty moss!

Gnome lichens hung out on rocks.
I'm likin' you!
I'm likin' you too!

And wood frogs chilled out in the icy pond.
Brr-ribbit!

Bog turtles cuddled in mud pies.

Grouses grouped under dormant grasses.

And skunk buddies nuzzled into their cozy den.

Meanwhile, the peat moss dispersed spores across the frosty bog.
Spread, little spores!
Fly like the wind!

When warm days returned,
along came a brand-new bunch
of bog babies!

spring
month 10

Beaver kits kidded in cattails.
Look at my tail!

Bog turtles paddled under pondweeds.
Did you see that bat?
Yes, yes I did.

And bumblebees tumbled among the blooming swamp flowers.
No thank you!
Why don't you tumble over here?

Woodcock hatchlings waddled in mud.

Otter pups pounced in puddles.

And skunk kits scampered in skunk cabbage

as wood duck chicks ducked inside their tree house.

Meanwhile, the soggy bog grew green again with blankets of fresh peat moss, welcoming . . .

heavy summer rains back
to the misty mountains.

summer
month 1

Where firefly friends fill the valleys.

Wood duck buddies whistle over rivers.

Welcome, rainstorm!

Runoff refills a mossy
mountain bog.

And boggy biodiversity abounds again.

wild turkey
Virginia
big-eared bat
ruffed grouse
bog lemming
river otter
water shrew
wood duck
wood frog
Carolina northern
flying squirrel
long-horned beetle
American woodcock

More about Bountiful Bogs

Bogs are a type of wetland that are typically formed in low-lying areas where water pools from slope runoff or groundwater seepage. These muddy habitats abound with biodiversity. Year-round, they provide a cool and moist home, nursery, and playground for all kinds of critters. This story features a meadowy mountain bog in the Blue Ridge Mountains. The groundwater-fed bogs, also called fens, found in these mountains are some of North America's tiniest and rarest habitats. Many of these small oases range from one-tenth of an acre to two acres in size. They are home to numerous threatened species, like carnivorous mountain sweet pitcher plants; four-toed salamanders; and North America's tiniest turtle, the bog turtle.

Bogs are covered in dense mounds of sphagnum moss, also known as peat moss. This moss soaks up water like a sponge, helping bogs control water levels during floods and refilling springs during droughts. Bog moss also cleans water as it filters through its leaves. And lastly, bog moss stores carbon from decaying plants. Peat bogs store twice as much carbon as all the forests in the world! In this way, bogs play an important role in cleaning Earth's atmosphere and in offsetting the damage from climate change.

Many bogs around the world have been drained, built over, or are drying out due to warming temperatures. But conservationists are starting to protect these rare and delicate habitats. Here are some preserves where you can see bog life in action: Bergen-Byron Swamp, New York; Brown's Lake Bog Preserve, Ohio; Cranberry Glades Botanical Area, West Virginia; Crescent Meadow in Sequoia National Park, California; Hawley Bog Preserve, Massachusetts; and Volo Bog State Natural Area, Illinois.

Author's Note

While I was in the process of making this book, Hurricane Helene swept through the southeast. I tracked the hurricane online as it traveled over some of the mountain bog areas that inspired this story. I worried about how these small wetland habitats would be impacted. Amazingly, according to the organization Conserving Carolina, the wetlands in North Carolina helped store more than one billion gallons of water during the hurricane! In doing so, they helped to reduce flooding and water pollution. They also offered refuge to wildlife during the storm.

Additional Reading

Sweeney, Linda Booth. *The Noisy Puddle: A Vernal Pool Through the Seasons.* Montreal: Owlkids, 2024.

Davids, Tracy, and Allison Cook. "Meet the Bog Turtle: The Tiniest Turtle in North America." Defenders of Wildlife, December 12, 2023. https://defenders.org/blog/2023/12/meet-bog-turtle-tiniest-turtle-north-america

"NPS Profile: What's In Our Wetlands?" National Park Service, updated March 15, 2021. https://www.nps.gov/grsm/learn/nature/wetlands.htm

Selected Sources

"All About Birds." Cornell Lab of Ornithology, n.d. https://www.allaboutbirds.org/news

"Bog Turtle (Glyptemys muhlenbergii)." The Nature Conservancy, May 11, 2020, updated March 11, 2024. https://www.nature.org/en-us/get-involved/how-to-help/animals-we-protect/bog-turtle

Ellis, Amber. "Exploring Mountain Bogs." The Appalachian Voice, August 10, 2014. https://appvoices.org/2014/08/10/exploring-mountain-bogs

Hess, Rachel. "Saving Our Rare and Beautiful Mountain Bogs." Conserving Carolina, July 31, 2020. https://conservingcarolina.org/saving-our-rare-beautiful-mountain-bogs

"Synchronous Fireflies." National Park Service, updated September 19, 2024. https://www.nps.gov/grsm/learn/nature/fireflies.htm

Schafale, M., R. Evans, et al. "Southern and Central Appalachian Bog and Fen." Nature Serve Explorer, n.d. https://explorer.natureserve.org/Taxon/ELEMENT_GLOBAL.2.723191/Southern_and_Central_Appalachian_Bog_and_Fen

BEACH LANE BOOKS • An imprint of Simon & Schuster Children's Publishing Division • 1230 Avenue of the Americas, New York, New York 10020 • Book design by Lauren Rille • • For information about special discounts for bulk purchases, please contact Simon & Schuster Special Sales at 1-866-506-1949 or business@simonandschuster.com. • Simon & Schuster strongly believes in freedom of expression and stands against censorship in all its forms. For more information, visit BooksBelong.com. • The Simon & Schuster Speakers Bureau can bring authors to your live event. For more information or to book an event, contact the Simon & Schuster Speakers Bureau at 1-866-248-3049 or visit our website at www.simonspeakers.com. • The text for this book was set in Fairplex. • The illustrations for this book were rendered in acrylic, marker, and pencil on Bristol paper and digitally collaged. • Manufactured in China • 1025 SCP • First Edition • 2 4 6 8 10 9 7 5 3 1 • CIP data for this book is available from the Library of Congress. • ISBN 9781665962674 • ISBN 9781665962681 (ebook)